AF632602

PRAISE FOR
WHEN FAIRYLAND LOST ITS MAGIC

'What a tremendous achievement this book is! Bijal and Rajiv make witty wonder out of awryness, find a story-telling tone and visual idiom that's somehow simultaneously grave, funny, furious and forwards-looking. They re-tell Grimm for our grim Anthropocene, but without (quite) letting hope slip out of sight. A subversive book that will make children want to read on, and also make them want to question the world beyond their windows and screens.' **ROBERT MACFARLANE**, author of *The Lost Words* and *The Lost Spells*

'Another gem by Bijal Vachharajani, and with added shine from Rajiv Eipe's fantastic illustrations. Rapunzel, Cinderella and other old friends bring us the realities of climate change, in a gentle yet powerful series of stories which all reading-ages will enjoy, read aloud, discuss and laugh and shake their heads over. It brings us even closer to the story people we love, because like us, they are suffering from the results of climate change. But in the end...no, wait for it.' **ZAI WHITAKER**, author of *Salim Mamoo and Me*

'All my life I've been disenchanted by fairy tales because they've all been unidimensional. And every time I've read them, I've always rooted for the villains; be it the child-eating witch from Hansel and Gretel or the Big Bad Wolf, because they've been the only interesting characters in the fairy tale world. Now, from the witch who wields her broom to make words flutter and the wizard who turns every surface he touches with his pen into gold, comes a series of fairy tales that I can actually enjoy, guffawing at the writing, and gushing at the illustrations, WITHOUT having to root for the villain: human-made climate change. Behind the grimmer-than-the-Grimms premise are two geniuses making merry with their own spins on fairy tales, and having the last laugh!' **ROHAN CHAKRAVARTY** *urf* Green Humour

ROYAL
BOUTIQUE राजा रानी बूटीक
BLACK FOREST
REAL ESTATE
STD ISD PCO
ZEROZ TYPING
ADVERTISING
PRINCE CHARMI
HAIRDRESSER
GRIMM & BROS
FAIRY GRANDMA
DAIRY PARLOU
ONCE UPON A TIME
WATCH REPAIR

WHEN FAIRYLAND LOST ITS MAGIC

BIJAL VACHHARAJANI

ILLUSTRATED BY RAJIV EIPE

HarperCollins*Children'sBooks*

First published in India in 2023 by HarperCollins *Children's Books*
An imprint of HarperCollins *Publishers* India

Building No 10, Tower A, 4th Floor, DLF Cyber City, Phase II, Gurugram – 122002

2 4 6 8 10 9 7 5 3 1

Text © Bijal Vachharajani 2023
Illustrations © Rajiv Eipe 2023

P-ISBN: 9789356995437
E-ISBN: 9789356995444

Bijal Vachharajani and Rajiv Eipe assert the moral right to be identified as author and illustrator of this work respectively.

The author acknowledges the quote from *10 Indian Champions Who Are Fighting to Save the Planet* by Bijal Vachharajani and Radha Rangarajan published by Duckbill Books in 2020.

The author acknowledges the quote from *Coraline* by Neil Gaiman published by Bloomsbury Publishing in 2016.

For sale in the Indian subcontinent only

All rights reserved. No part of this publication may be reproduced, stored in a retrieval system, or transmitted, in any form or by any means, electronic, mechanical, photocopying, recording or otherwise, without the prior permission of the publishers.

Cover design: Maithili Doshi
Cover illustration: Rajiv Eipe

Book design: Studio Em & En
Typeset in Capitolium2

Printed and bound at
Nutech Print Services - India

HarperCollinsIn

'Climate change is an amorphous fire-breathing dragon that exists out there, but we can't see it.'

—**ROHAN ARTHUR**, Scientist, in an interview for *10 Indian Champions Who Are Fighting to Save the Planet*

'Fairy tales are more than true: not because they tell us that dragons exist, but because they tell us that dragons can be beaten.'

—**NEIL GAIMAN**, Author, *Coraline* (paraphrasing GK Chesterton)

In a land far, far away,
a book opens.

The story begins,
like many stories do...

ONCE UPON A TIME

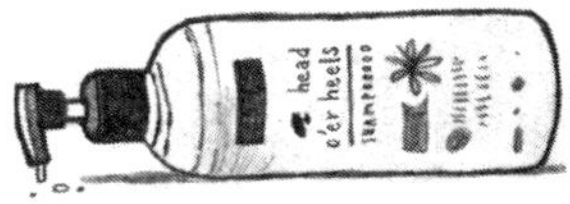

Rapunzel's Hair Washing Day

Once upon a time, a man broke into a garden to steal a sprig of rapunzel for his wife. While he was busy clipping the leaves, he was caught by a sorceress who, of course, owned the garden. She, of course, retaliated by stealing this couple's child and locking her up in a tower so high that it rivalled most industrialists' fancy buildings. The girl grew up to be Rapunzel with hair so long that it rivalled the number of zeroes in said industrialists' bank accounts.

But Rapunzel was having a tough time maintaining her hair, especially since the Great Water Crisis had hit Fairyland. Now, the princess could only wash her hair once a week, on Thursday

between 6 a.m. and 6.25 a.m. when free-flowing water came in the tap. One such Thursday, she washed her hair and, at 6.27 a.m., opened the window of her tall tower so that she could let her hair down to dry and for a handy prince to use as a ladder. Rapunzel retched; she coughed, her eyes watered. The air outside was SO yuck! She hastily shut the window and glowered at the grey smog whirling outside her tower. Meanwhile, the prince kept waiting and waiting, but there was nary a sight of hair or hare.

No White and Her Seven Friends

Once upon a time, a queen was busy sewing clothes when she pricked her finger. Three drops of blood fell on the snow outside—and she wished herself a daughter whom she'd call Snow White. After many years, her child was born. But, by then, there was no snow in their part of Fairyland, so the baby had been called No White.

No White had luxurious black hair just like ebony, skin white as snow and lips red as blood, which were unfortunately hidden behind a mask. Threatened by a Stepma—long story, not important—No White left her home and moved in with seven friends. They used to have real names but ever since they began working at the coal mines, their names had been changed to Carbon, Methane, Nitrous, Oxide, Water, Vapour, and Gas. It was unfortunate, but what to do?

Carbon was no longer happy.

Methane was sneezing even more because of all the pollution.

Dr Nitrous had so many patients at the factory, all down with a weird flu, that he had no time for anything else.

Oxide had curled up and gone to sleep.

Water was constantly roiling and boiling, just like the rising sea levels, and he was even grumpier than usual. Who knew that was even possible?

Only Vapour and Gas weren't anxious. They went around singing la-la-la. When the smog overwhelmed their lungs, they

stopped and pretended like they couldn't be bothered to sing. 'Smog, what smog?' they gasped.

'Nothing is wrong,' Gas wheezed.

'It's all fine,' Vapour said with red, watery eyes.

But every smoggy cloud has a silver lining. No White discovered that she did not need rescuing from her Stepma. In fact, nature had solved her problem. She was most amused as Stepma's plans to poison her with apples bore no fruit. Fairyland was witnessing an apple scarcity because the rains had come too early and damaged most of the harvest. Then a blight had finished off the rest. The last anyone had heard of Stepma was that she was trying to inject poison into a jamun, but the fruit kept staining her hands purple. Even Mirror, Mirror on the Wall was horrified at her purple fingers. No White laughed and laughed a tinkly laugh, which at any other time would have attracted birds and butterflies. Now there were none for her to play with.

Meanwhile, the poor prince kept waiting for No White to bite into a fruit, but there was no sight of apple or jamun.

gasp!

The Not-Sleeping Beauty

Once upon a time there was a princess.

A very beautiful princess.

A very beautiful princess, beloved of her parents, and her subjects, and objects.

But alas, she was cursed. Cursed to fall into an almost-death-like sleep. Waiting for a handsome prince, who would never arrive because...

The curse would not take! The princess in question—Sleeping Beauty (SB)—wouldn't sleep. Which left her simply as Beauty—and, really, that was a name that simply would not do.

So far, the insomniac princess had tried many things to sleep—a cup of haldi doodh (sprinkled with a kiss of nutmeg), a long hot bath, a short cold bath, lavender candles, lavender cream, lavender tea, and even gross banana tea. All it had done was make SB full of calcium and feel very clean.

The reason SB could not sleep, despite cursed spindles, was

that she was feeling quite anxious. Every time SB spun her cursed spindle, and pricked herself, she would fall down dramatically on the floor and wait.

And wait.

And wait.

And nothing—still no sleep.

On the hard floor of the Sleeper and the Spindle Room, she would think and think, and worry and worry, and cry and wail, all of which made her anxiety even worse.

It had become so bad that the King and Queen were getting quite anxious too. If she didn't sleep, then when would the handsome prince come and kiss SB and wake her up? The poor prince also wanted to know; he kept waiting, but there was no sign of a dream or a snore.

This was a very serious issue. Her parents consulted the old fairy—the good-turned-evil-turned-good-again one—who had cursed SB's family after they ignored her social media post wishing them happiness on having a baby girl. But curses, once given, could not be returned.

Everyone kept telling her everything was fine. SB knew that things weren't all 'fine'. She was filled with dread—something her therapist had diagnosed as climate anxiety. Instead of dream clouds, a permanent black cloud hovered over her.

Goldilocks, Payasam, and the Three Bears

Once upon a time, Goldilocks stole into a den where lived Three Bears. And she had been visiting ever since. She couldn't help it; the piping hot, golden payasam, flecked with jaggery and cardamom, was irresistible. Yum!

Goldilocks sat on the first big chair, and tried a spoonful. 'It's TOO hot!'

Then she got up and sat on the second big chair. She sniffed the payasam, mmmmm. But again, the spoilt brat pronounced, 'It's TOO cold!'

Goldilocks moved to the just right chair.

'And that one's perfect, no?' Mama Bear pointed at the last dish. She was beaming fondly at Goldilocks.

But then, something horrible happened. Goldilocks tossed

?
wail!

back her golden locks and wailed. 'IT'S TOO HOT! IT'S TOO COLD!' How was that even possible?

Papa Bear frowned. He pushed Little Bear in front and clicked his tongue, 'Try it, baba, try it.'

Little Bear put her little paws around the bowl and tried the payasam. 'IT'S TOO HOT AND TOO COLD!' Little Bear wailed.

Mama Bear shook her head and growled, 'Nothing is as is.'

I

NOTHING IS AS IS

Once upon a time, nothing was as is in Fairyland. Mama Bear was right. Their beloved homeland was turning into the Land with Strange Weather, the Land of Smog and Dust, the Land with Too Much Snow (where it wasn't meant to be snowing and where No White definitely did not live) and Too Much Heat (where it was meant to be cold).

The glaciers were melting at such an alarming rate that *The Fairyland Times* had called it the worst climate crisis since the Narnian Ice Age.

Unseasonal thunderstorms had never been uncommon in Fairyland—they happened every time the Snow Queen or some other high-and-mighty being was feeling angry. But now, it was such an everyday occurrence that it felt like all the high-and-mighty beings were in a perpetual, collective foul mood.

Rising sea levels meant that everyone was constantly bobbing about in boats to get from one castle to another. The people in charge of boats were levying all sorts of exorbitant prices to ferry people across—a favourite teddy bear, an eye, a soul. After all, it was a perilous journey, the moat monsters now had free run of the waters, and no one knew when they would become, like a certain actress in a certain Bollywood film, a chew toy for the Alley-Gators.

Something had caused the weather to turn fouler than anything the Evilest of Wizards and Witches could conjure. Curses were not working. The morals of Fairyland's stories were messed up. And there were no Happily-Ever-After sundaes because chocolate was so rare that only royalty could afford it.

No wonder, in the now Disenchanted Not-a-Forest, Big Bad Wolf was feeling quite sorry for himself. He was genuinely tired of people spotting him easily and running away, yelling 'Wolf, Wolf'. Including that Hunter bloke, who was obsessed with science stuff. When not poaching Bambi's Amma or some other animal, he was watching some documentary about the natural world and, before he raised his bow and arrow, he'd reel off trivia, such as

'Roe Deer: almost tailless, reddish brown in summer and greyish brown in winter', and then let off his arrow. It was most annoying. It was even more annoying when the Hunter bloke saw Wolf and yelled, '*Canis lupus, Canis lupus*', before tripping over his own big feet.

Hello! Wolf knew who he was, and he didn't need anyone telling him his species at the top of their voices. If they wanted him to have a name, he'd pick something like, say, Gandalf the Grey. He didn't look at a human girl and say, 'Oye, *Homo sapien*'. He said, 'Oye, Red Riding Hood'. Because that was her name, for reasons best known to her family, who seemed to think it was appropriate to name their child after the clothes they wore. And really, who wears bright red clothes while walking in a forest? It was like she was yelling to all predators to come gobble her up.

Not that he had seen much of her to be able to gobble her up recently.

The once-Enchanted Forest had been chopped down by all the loggers, and the animals hunted or driven away by the hunters. Who would want to live in a Disenchanted Not-a-Forest? Where once stood magical baobabs and banyans, brightly painted flame-of-the-forest and amaltas, and mighty oaks and sequoias; there was now a land full of tree stubs and palm trees. Apparently palm oil was very profitable. But now everything looked the same: BORING,

with a capital B-O-R-I-N and G. Birds had fled, so no one twittered around the princesses. The bees and wasps were confused, so they didn't really know what to pollinate. Hot winds blew away the spiders and brought with them locusts instead.

Result: deer and other juicy food species had moved further away. Just like the Three Pigs, who, rumour had it, had migrated to the city in the Fairyland district to work in construction. Apparently, they were in high demand for their skills in building houses with straw, sticks, and bricks.

And with no mighty trees to hide behind, Big Bad Wolf was immediately spotted by all and sundry, who ran away yelling 'Wolf' or '*Canis lupus*'. Result: Big Bad Wolf was constantly hungry. This was the second reason he felt sorry for himself. He didn't know how to scale palm trees to get to date palms. He wouldn't mind a juicy fig, but he didn't want dates. Not that he'd be caught dead eating them. He was not going to turn vegetarian, no way. He used to be Big Bad Wolf of Fairyland. The Terror of Pigs. The Huffer and Puffer of Houses. The Gobbler of Grandmas. Now, he was simply known as Big Wolf; even the Bad had been dropped.

On top of all his woes, there was a ringing in his Big Ears; it wouldn't go away. It sounded like a wind blowing through a tunnel. He couldn't think clearly because of it. He let out a loud howl to

WOLF!
canis lupus!

drown out the noise and settled down on the Not-a-Forest floor. It was hard and uncomfortable, given it hadn't rained in his part of the forest in forever. His belly growled in sync with his howls. He licked his grey fur, to console himself. 'What is the point of my Deep Voice, Big Eyes, Big Hands and Big Mouth if I have nothing to eat?

This has to be someone's fault,' he thought. 'Someone's got to know what's happening. I know what to do!' Filled with purpose, if not porpoise or any other meat, Big Wolf sprang up and raced to the Snow Queen's palace. After all, who better than their Queen of all Queens, the one with snowy powers, to have the answers?

2

A MER PROBLEM

As he ran, Big Wolf heard someone crying. He cleared his Big Ears. The windy sound was still there, and now there was another horrible sound. As if someone had taken ten cats and was forcing them to behave like dogs, making the cats wail. Curious, Big Wolf turned left and skidded to a halt.

In front of him was a very familiar cottage. The horrible wail was definitely coming from inside. Big Wolf sniffed—he could smell gingerbread, which is what the cottage wall was made of. Cautiously, he inched towards the sparkling sugar window, careful to crouch low. He looked inside and his whiskers quivered.

There was the witch. Storybooks really exaggerated her features—her hooked nose, jowly jowls, and creased forehead. Big Wolf quite liked her. Sometimes, if she had caught many children, she'd invite him for brunch, complete with leftovers and a Bloody Mary mocktail.

Nice, Nasty Witch, he smiled. His stomach rumbled, and he thought of going in and seeing if she was cooking anyone fresh. Then he realised she was sitting by her oven and weeping. He sniffed the air again. Something was missing amidst the cloying smell of stale cake and crumbling pastry. Uh-oh, he realised, there were no children being roasted in the oven. It was as empty as Big Wolf's belly.

Big Wolf thought of going in, but he first checked her social media—she had just posted a picture of the empty oven and a caption that said: not captured a single child #sad #hangry

Hanging out with a hungry, angry Nice, Nasty Witch was never a good idea, so Big Wolf decided instead to seek counsel from his BFF, Little Mermaid. It was a bit of a detour, to be honest, stopping by the Ocean of the Underwater Kingdom, which was on the other side of his Disenchanted Not-a-Forest. But if he was lucky, Mer would have some tasty fish for him. Licking his Big Sharp Teeth, Big Wolf changed course. But it seemed that today

was just not his day. When he got there, Mer was in quite a flap. She was in the tide pool, giant tears adding to the salty water. An octopus hastily passed some kelp to Mer to wipe away her tears.

'Aiyo, what happened?' Big Wolf asked Mer. 'That Prince again?' He felt terrible seeing her so upset. Ever since that Prince had been rescued by Little Mermaid, she had transformed into a Boring Version of Herself. Constantly talking about having two legs. It made Big Wolf growl; two legs were so useless. He was either Team Four Legs or Team Fin, Fur, Feather, and Tail.

Of course, truth be told, it was a tricky friendship in many ways. Mer was wickedly funny, making her fun to hang out with, but she also smelled so good. Like his favourite fish fingers. He'd finally understood what that book *Twilight* was all about. But they were supposed to be friends, or so she had made him promise. They had even done a pinkie promise. So, he could not eat her. Yet. Also, she was the only person—okay, merperson—who did not turn tail at the sight of him.

Big Wolf stood there awkwardly as Mer wailed and flapped her tail, splashing him with water. He shook the water from his tail and gave her a lick. 'Why are you adding so much salt to the ocean? Highly unnecessary, I tell you.' The ocean was already turning acidic, and didn't need to be any saltier.

'I went on a field trip today,' Mer said and freshly burst into tears. This was very perplexing to Big Wolf—wasn't a field trip supposed to be fun? Where you got to learn things, or see new things, and eat a picnic dabba. Big Wolf started salivating at the thought of fish and chips, but he firmly forced himself back to his friend's problems.

'So...?'

'All the corals are WHITE! WHITE AS GHOSTS. WHITER THAN WHAT THE LAUNDRY ADS PROMISE. WHITER THAN SNOW WHITE! I mean, No White.'

'So...?'

Mer glared at Big Wolf. The ocean churned menacingly, the waves suddenly looked bigger and stormier. He swallowed and added hastily, 'I mean maybe they are just, you know, falling for those fairness cream ads we keep seeing?'

Despite herself, Mer giggled. 'You're truly a blistering barnacle.' But then she looked sad again. 'Don't you get it? The ocean is warming. The first sign is always the corals turning white.'

A warming ocean really did not sound all that bad to Big Wolf, who was suddenly imagining seared striped bass, seasoned with the salt from Mer's tears. Delicious! He started salivating again.

'Oh, you're impossible to talk to when you're hungry,' Mer said. 'Hang on.' She closed her eyes and reached out over the water. The

ocean churned and spat out a dead fish. 'Here, snuck it from the Old Fisherman's tattered net.'

Big Wolf would have hugged his friend, but he was so hungry he might have just eaten her up. First, the fish. Then, a chat about his Disenchanted Not-a-Forest, the noise in his ears, and these fair-and-not-lovely corals. And then, off to the Snow Queen, he decided. Between Mer and him, they could ask more of the fairy people, who stayed on land and in water, to join them. The more the merrier, and mightier. Some use of those Twittering Trolls who otherwise just spread rumours.

But first, the fish. Needs must.

At the same time, SB was pacing in her room. In such a frenzy that she didn't hear the knock on the door or see No White slip in.

No White looked around the room—it was meant for a sleeping person. A bouncy mattress, fluffed-up pillows, drapes that didn't allow any light to filter in, and a night lamp shaped like an owl. She sighed and sat next to her best friend forever.

'Still nothing, I can see,' she said, removing her mask. 'Oh no.'

SB held up her hand—all ten fingers were bandaged heavily from her constant pricking by the cursed spindle. She shook her head and sighed.

No White slid her phone out of her pocket and summoned a Twittering Troll. Hulky Twittering Trolls were a new addition to Fairyland. As messenger birds flew further away in search of food, these club-sporting trolls were put to task to share messages between queendoms. They spread news and even rumours.

And all they wanted in return was to be fed more juicy stories about fairy people.

'I have heard something...' No White's voice dropped to a whisper even though no one else was around. 'You won't believe it. I think people need to know about it.'

3

CINDERELLA DOES NOT HAVE A BALL

Unlike everyone else in Fairyland, things were going great for Cinderella. From growing up with her stepfamily, covered in cinders, to ruling over her own little pocket of queendom. Alas, the Prince had died. A tragedy, really. He'd tripped on one of her blue glass slippers, tumbled straight out of the palace window, and, well, that had been that. So much for happily ever after. Social media had been abuzz with gossip for months after.

PAH!

But Cinderella was happy. She hadn't particularly wanted to marry the Prince. Or actually marry at all. But what was one to do but go with the flow? He hadn't been a bad husband, just a little vain and flighty. C'mon, who marries a woman after one dance? And what if he had met other women with tiny feet before her?

But no complaints. Cinderella had her Fairy Nani, who had retired from magic, except for a few spells here and there to enchant Cinderella when she was bored. Mice stayed mice. And pumpkins used to be pumpkins, to be turned into sabjis or soups. And she was no longer in danger of reliving that humiliating moment when she found herself sitting on a pumpkin at the stroke of midnight.

Never again. Mainly because pumpkins had gone extinct.

Cinderella brushed her ashy hair for the 100th time and smiled at herself in the mirror. 'Looking good,' she winked at herself. As she wafted down to her throne room, a page came running and said, 'Your Highness, Your Highness. Did you hear?' She stopped and pointed at the window.

Cinderella raised an eyebrow at the page. 'Page 42, calm down! Where's the forest fire?' She smiled at the girl kindly.

Page 42 swallowed nervously. 'Umm, Your Highness, there's a Twittering Troll here, for you.' She again pointed at the window.

'Who is the troll from?' Cinderella snapped. All that time by the fireplace in her father's house had made Cinderella have a quick, hot temper. It rose as quickly as it fell.

HA HA HA
HA!
tee hee
!
HA HAA!

'It's from, oh, it's from...' The page twisted her hands in alarm. Cinderella turned towards the window and scowled at the Twittering Troll.

He waved his club, and then stared dully as it slipped out of his warty hands. He dug into his ear and retrieved a gross-looking waxy scroll. 'Do you think,' he grunted. 'Cinderella has something to do with all this changing climate? #FossilFuel #Cinder=Carbon, get it, get it? LOL.' It was from No White.

The troll belched loudly, picking up his club and thumping away, leaving a seething Cinderella behind.

'How dare she,' Cinderella cried, her voice rising along with her temper. Her green eyes flashed with anger. 'How dare she!' She stomped into her throne room, glared at the sentries and ministers and swooshed up to her throne. 'You...' she said, pointing at one of the sentries. 'Get Fairy Nani. NOW!'

Two sentries ran out faster than the others.

Fairy Nani flew in two minutes, six seconds later, looking displeased. She had been watching a rerun of her favourite show, which involved six friends and a coffee shop, and just as one of them had ordered pizza with anchovies, she had been summoned. Grumpily, she stared at Cinderella over her spectacles, and her translucent blue-silver wings fluttered menacingly as she folded

Do you think
Cinderella has
something to
do with all this
changing
climate?
#Fossil Fuel
#Cinder=Carbon,
get it, get it?
LOL

her arms over her soft belly. 'What now, Cinderella? Why are you angry…' She didn't add 'again' but she muttered it under her breath.

'THEY'RE SAYING I'M RESPONSIBLE!'

'You are quite responsible,' Fairy Nani replied, looking perplexed. 'It's kind of obvious; you've managed your queendom so well. Why would it make you angry? One must always accept the truth with dignity.'

'ARGH,' Cinderella shook her head. 'Responsible for this… this climate.'

'Pardon?'

Fairy Nani took out her wand—also blue and silver—and conjured up a steaming mug of hot chocolate. 'Drink that.'

The mug of hot chocolate went whooshing to Cinderella, who caught it and promptly dropped it. 'Ouch! It's hot.'

With a swish of the wand, the mess was mopped up, and another mug handed to Cinderella. 'Drink, drink.' Fairy Nani frowned at the colossal waste of the precious chocolate.

Cinderella took a sip of the creamy hot chocolate. She felt her insides unclench. Her face was no longer as red.

'Now, tell me what happened?' Fairy Nani asked.

'That No White! She trolled that I am responsible for this turning of the climate, because of... of... my name. Why didn't she eat that poisoned apple when apples were still around? Would have done us all a favour.'

Fairy Nani stared at her ward and a wave of sadness overcame her. This sort of fake news was so damaging. 'I did tell you to stick to Ella,' she pointed out, fluttering on top of a chandelier and making herself comfortable there. She muttered 'I told you so' under her breath.

'But everyone knew me as Cinderella! How could I... how could I go back to just plain old Ella? And everyone kept teasing me with that song, "Under my umbrella", Ella.'

Fairy Nani sighed loudly as she gently swung. 'Well, you're just going to have to clarify that you are not emitting carbon just because you're nicknamed Cinder.'

'No one's going to believe me. No one! You should have heard the Troll. No White just wants to blame someone. And I am the perfect blamee.'

'You mean victim,' Fairy Nani tutted. 'What is blamee? Don't talk rubbish.' Under her breath she said, 'It's not like you live a completely carbon-neutral lifestyle; jet-setting to meet Rapunzel one day, and buying so many unnecessary things all the time, all those dragons being bred by the acres, that non-stop mining and opening carbon-fuelled industries, and what nots.'

Cinderella put down her mug and collapsed on her throne, in tears. She looked so sad that Fairy Nani started weeping as well. When she saw her favourite (read: only) godmother crying, Cinderella's temper fizzled out. 'Chin up, chin up,' Cinderella said. 'Tell you what, we will go to the Queen Who Unites Us All, Snow Queen, and she will set everyone straight.'

Fairy Nani immediately cheered up; she swished her wand and sprinkled fairy dust around the throne room as she fluttered out of the door. Page 49 sighed silently. Cleaning fairy dust was just the worst.

sigh!

SNOW QUEEN'S
PALACE

4

WATCH OUT FOR THAT PUDDLE

By the time Cinderella's cabbage carriage rode up the mountain that led to the Snow Queen's palace, Fairy Nani's tears had dried up. It might have been the heat evaporating the tears—strange, given they were now way up in the North Pole of Fairyland.

'Chin up, chin up,' Cinderella murmured for the 678th time. She tapped Fairy Nani smartly on the knee and said, 'Look we're almost here.' She pointed in the direction of the lake, the Mirror of Reason.

Fairy Nani, whose hands had been on her face all this while, looked out from behind them and exclaimed, 'Withering wings! What happened here?' Cinderella turned to where she was pointing. The last time they had visited the Snow Queen, the frozen lake had been stretched out like a big mirror, reflecting the sky, the palace and everything around it. Fairies could be found fluttering on top of it, admiring their own reflections and pulling faces at the frogs who used to be princes.

But now, the lake was not frozen at all! Forget brain-freeze level of frozen; it was not even half-frozen like a melting kalakhata gola! It was running on empty, with small sad patches of water here and there. A throne bobbed in one of those puddles.

'OH NO!' Fairy Nani wailed.

The Puddle of Reason

'Now what?' Cinderella asked.

'They will blame this one on you as well—that you came and melted the Mirror of Reason.'

Having spent three days and a quarter in that tiny carriage with a weeping willow of a Fairy Nani, Cinderella's already slim patience was hanging by a frayed thread. She poked Fairy Nani with her own wand. Sparks flew out. 'Be reasonable. I just arrived. This lake has been melting for a while. This change does not happen overnight.'

Fairy Nani cheered up at the thought and looked out once more after retrieving her wand. But again, she burst into tears.

'What now?'

'Look!' Fairy Nani pointed. They had just turned a corner, and—

A long line of carriages, carts, steeds, dragons, and fairy folks were making their way across the mountain. Goblins and dwarves, giants and Twittering Trolls, bears and children, witches and wizards, elves and fairies—were all there, some with packed suitcases and bundles, some with rucksacks, and some with just garbage bags stuffed with their belongings. Many of them looked like they had left their homes forever.

'What is happening?' Cinderella asked.

'Can't you see?' A big furry grey snout poked into the carriage's window.

Cinderella let out a loud yelp, quite unlike one expected of a princess. Fairy Nani fell off the seat, her wand casting a spell by mistake. White mice ran out of the tip of her wand and the furry grey snout opened and swallowed them up. 'Yum, thank you!'

'Oh, Big Wolf, you really gave us quite a fright,' Cinderella said, recognising the voice.

'Huh, me?' Big Wolf was running alongside the carriage

effortlessly. After all, he had such big legs, all the better to run with. 'Mer and I spread the word; we told the fairy people that we all needed to rally together and talk to the Snow Queen.' He stopped in his tracks and yelled, 'Look up!'

Cinderella and Fairy Nani poked their heads out of the carriage and looked up. 'Oooh look,' Fairy Nani cried.

'Oh,' Cinderella sounded sad. 'You can barely see it anymore.'

'No, no, I can see very clearly,' Fairy Nani said, waving her hand left and right slowly, like the royals do. 'Helloooo! Princess Chameli, here!' She waved again.

Princess Chameli waved back. There seemed to be more people on the carpet, but they were obscured by the smog.

'Not the magic carpet,' Cinderella said, rather crossly. 'The Northern Lights.'

Fairy Nani squinted up. 'Where?'

'Exactly!' Big Wolf sounded rather pleased and perplexed at the same time. Pleased to be proven right, perplexed because the Northern Lights had DISAPPEARED!

The Northern Lights were the entry point to the Snow Queen's palace. They shone purple and green, blue and silver, pink and red—like someone had shaken rainbows in a bottle full of fireflies, and then flung the mixture across the entire sky (do not try this at

home). It was a majestic sight that made human children believe in magic. But now, the night sky was dark, as if someone had extinguished the lights to save energy.

Instead of an endless, shimmering diamond sky, it was a menacing, inky smog-filled, purple one.

5

THE THAWING OF THE SNOW QUEEN

It took another day and a half for the cabbage carriage carting Cinderella and Fairy Nani to reach their destination. By then, Big Wolf had already reached, ahead by 4 minutes and 57 seconds. By then he was panting and pointing in turn.

'That… look… oh no, what happened!'

Immediately, Fairy Nani burst into tears.

Cinderella left her in the carriage and jumped out, slamming the door behind her.

'What now?'

'There!'

'Where?'

'Right there!' Big Wolf snapped his Big Sharp Teeth. 'Can't you see?'

'But I can't!' Cinderella said, fully puzzled.

'Exactly.'

Because there was nothing to be seen. No palace of drifted snow. No doors and windows made of cutting winds. No snow-blown rooms behind those no-doors, and no windows to let in the lights of the Aurora.

'Where's the Snow Queen's palace?'

Big Wolf pointed at a puddle.

'A puddle? How is the palace a puddle?' Cinderella was all eyes and mouth. Where once stood the cold palace, there was now a puddle and debris of paintings, clothes, jewellery, beds, sofas, bobbing about in wet patches. In the centre, on a piece of driftwood, sat the Snow Queen.

Usually, the Snow Queen was one of the most awe-inspiring sights. She was like an ice sculpture made by the finest Renaissance sculptor ever—her chiselled features, her translucent icy skin, and her glacial blue eyes that shone like stars in a night sky. She looked like she was made of a trillion snowflakes spun with fireflies.

But now, she looked like an ice sculpture that had been left on top of a hot oven—a little melty at the edges, tired and weary.

And, to be honest, a little bit of a drip. If anyone had been feeling reckless, they'd have dubbed her No Queen.

Cinderella and Big Wolf approached her tentatively, hearts thudding fast. They curtsied and then stood. And stood. And stood. The Snow Queen didn't even look up.

Cinderella nudged Big Wolf. 'You talk to her.'

'Arre, what is this behaviour? You talk to her. You're a princess.'

'But you are Big Wolf, and you know… ummm… with a Big Mouth, all the better to talk with,' Cinderella told him with an impressively impassive face.

Big Wolf stuck his Big Tongue out at her and said, 'Fine, use my story on me, won't you!' He turned to the Snow Queen and let out a big howl.

The Snow Queen stirred as if being roused from a deep sleep. 'My snow bees,' she said. 'They disappeared one day. Just gone.'

'Just like the bees in our forest,' Big Wolf growled.

'And my power…' She slumped back on her piece of driftwood. It was like she was frozen atop it, which was a bit unlikely, given it was at least fifty-five degrees right now. 'It disappeared. I tried everything… everything to bring it back. But it seems to have melted away, just like my castle. Gone. GONE! GONE LIKE THE GLACIERS.'

By then, Fairy Nani had calmed down and come out of the carriage, dabbing her eyes with a sparkly lace hanky. She swished her wand and conjured up squishy sofas for them to sit on.

Big Wolf whimpered. Fairy Nani sighed, swished and drew him a squishy bed. He sniffed and settled down, sighing with relief. He had had a long journey. And he didn't have cabbage carriages like some posh princesses.

'How long has she been like this?' Cinderella pointed at their queen. She looked towards the fairy people; everyone shook their heads or shrugged.

'We just arrived,' Mama Bear said. 'Got your message, Big Wolf. Thanks.' Papa Bear was busy opening tiffin boxes and spooning payasam into Goldilocks and Little Bear's mouths. Nice, Nasty Witch was right behind them.

A voice piped up.

'For too long, oh woe is me.

Even longer than the song, doe re me,'

It was the Golden Stag aka Ranga Rhyming Reindeer (RRR), who had come and settled next to Big Wolf. Suddenly spotting him, they sprang up.

'I'm endangered, so don't you dare eat me.

Am now officially a protected species, see?'

They scuttled over to the sofa instead, as Big Wolf looked on dolefully. He wouldn't have eaten the reindeer, at least not more than a few bites. It's not like they would miss some meat off their flank. They had plenty of it. Big Wolf sniffed delicately.

RRR sighed and shook their antlers.

'Nothing has worked, nothing at all, not even a bit.
We've tried everything, but she won't snap out of it.
People are forced to move, animals are losing their home.
It hasn't spared anyone, nor fairy, nor goose, nor gnome.
Forest fires run amok, and our fire fairies can't put them out.
Floods and droughts everywhere, all we do is shout or pout.
Something must be done, yes it must.
Or Fairyland will just go bust.
What to do now? I ask you all.
Have you ever heard a glacier fall?'

Big Wolf wanted to say he lived in the Disenchanted Not-a-Forest, a tropical area of Fairyland, so how was he going to see a glacier, let alone hear it falling? But one look at RRR's antlers, and he decided to hold his Big Tongue. Those antlers were rather sharp and pointy.

Cinderella shook her head.

'It's a terrible, terrible thing. Rips you and leaves a black hole.

The ice cracks, a booming sound that shakes your sole.' They pointed at their hoof.

'Ummm, you mean soul?' asked Big Wolf. 'Ooh, or sole, like the fish.' Big Wolf began to salivate at the thought of a sole.

RRR looked at him mournfully.

'You feel the sound in your feet. It's hoof-clenching.

And yes, fine, also in your soul. It's heart-wrenching.

This mighty glacier, freezing our stories, oh re,

Crumbles memories, identities, ice stories swept away.'

'Ice… story?' Big Wolf hastily looked away as Cinderella shook her head and frowned at him.

'Oh yes, the ice tells stories,' another voice added. 'Of ages gone by, of deeds and actions, of the way our world has changed.'

Big Wolf was tired of new voices chiming in. Already he had that windy sound in his ears, and then new voices kept being added to this crowd.

This time it was Gretel and Hansel. The two children had grown since they had been last spotted in the forest, evading his friend—Nice, Nasty Witch. They were now teenagers, he realised. It was hard to tell the siblings apart, both of them wore almost identical clothes—brown shorts and trousers, green shirts, and a forest green jacket each. The colours made him miss his forest. Their brown hair too had grown long, and were now pulled back into ponytails. They really looked like children of the Disenchanted Not-a-Forest.

'How do you know this? All this story business?' Cinderella asked, a tad scornfully. She was not a big fan of those two siblings. Always playing together, never fighting. Refusing to be someone's meal. Terrible behaviour. Characters should do what they're supposed to do, believed Cinderella—witches casting spells, children getting eaten, and so on and so forth.

'We learned it in school,' Gretel said. 'And in books. And now, we're here to talk to the Snow Queen. And maybe help.'

Fairy Nani and Cinderella rolled their eyes. Behind them, the fairy people who had come to meet the Snow Queen also rolled their eyes.

Only Big Wolf didn't. Not because he didn't have Big Eyes to roll.

He'd have been great at it. In fact, he'd totally win an eye-rolling contest any day. He didn't because the siblings kind of looked like they knew what they were talking about. Also, so far no one else had offered to help. And no one was really clocking on to the fact that the Snow Queen wasn't in a state to help either. So, while he was grasping at (paper) straws, at least there were straws to grasp.

Gretel and Hansel walked up to the Snow Queen.

'Snow Queen,' Hansel said. The monarch didn't stir. Gretel and Hansel looked at each other and nodded determinedly.

'All right, enough,' Gretel said. 'Stop stewing and moaning and whining.'

A collective gasp went up in the air. 'The cheek!'

'These children, no manners.'

'Nice, Nasty Witch should gobble them up. Where is she when you need her? Turns up every time I am making aloo luchi, of course.'

'Their poor parents would be so embarrassed.'

The siblings didn't care. They crossed their arms and stood in front of the queen. Big Wolf reluctantly got up, went and stood by them, crossing his Big Arms as well.

Shocked, the Snow Queen snapped up straight on her driftwood throne. 'How... how... how dare...' and then she threw her arms up in the air, and said, 'Oh you're right, absolutely right. I need to pull myself together.'

Another collective gasp went up in the air.

'So brave.'

'These children, such stars.'

'I will slay the Nice, Nasty Witch if she tries to gobble them up. Watch. Will stuff and cook her in my luchi only.'

'Their parents must be so proud.'

6

THE FINAL SHOWDOWN

As the hubbub died down around the Snow Queen and the two once-upon-a-time-impudent children, suddenly…

'You?'

Big Wolf turned again, tiredly—at least this time he knew the voice. It was Cinderella. And she sounded like she had swallowed a whole swarm of snow bees. She was pointing into the crowd. He squinted.

Out of the sea of fairy folks emerged nine people: No White, SB, Carbon, Methane, Nitrous, Oxide, Water, Vapour, and Gas. But Cinderella was pointing at No White. 'You, you, you!'

No White trilled back, 'Me, me, me, me?' And promptly burst into giggles.

That only made Cinderella redder in the face. She sprang up from her squishy sofa, or at least she tried to. It was so squishy and squashy that she fell back. She ended up wriggling to the edge of the sofa and then jumping up. 'You dare show your face, after… after…'

No White suddenly turned paler than usual. She blinked rapidly and swallowed. 'Ummm, you heard that troll, did you? Oh no.'

She looked nervously at SB, who whispered to her friend, 'I told you not to send that troll.'

'Heard it, and saw it getting trolled around Fairyland,' Fairy Nani hollered from behind. 'I mean I didn't hear it, but I heard that everyone heard it. I am not on social media, you see. Have better things to do with my time.' Saying that, she beamed at everyone and sprinkled fairy dust all around. Everyone started sneezing and coughing and quickly moved away from her.

Cinderella crossed her arms just like Gretel and Hansel had, and stood there tapping her feet.

'I… I…' No White spluttered. 'I didn't think I'd see you until the Re-enactment of the Day Cinderella Met Prince. So, you know I did…' She stared at Cinderella for a minute and then spat back, 'Look, it's not like your name is not Cinderella. It is. And that just shows.' No White stamped her foot.

'What does it show?' Cinderella said, her voice in a scary whisper, which made Fairy Nani think of the Godfather.

'It shows that you're behind all this. You! Cinder-Ella. That's you.'

SB tried to hold her friend's hand to make her stop. 'Enough, really, No White.'

Cinderella glared at No White and said, 'Then what about your Greenhouse Gases named friends?'

No White turned even paler. 'How dare you!' she cried. She stamped her foot. Once. Twice. Thrice. 'GHGs are crucial to Fairyland's atmosphere. We all know that. What we don't need is burning fossil fuels rampantly. Missed your geography lesson, Cinder?'

Big Wolf growled, 'They are the ones working in mines because their forest has been dug up for its heart of gold. And, they get affected by the deplorable conditions.'

Just then, Gretel walked up to them and said, 'What's in a name?'

Cinderella frowned.

As did No White.

As did SB and Big Wolf.

As did Carbon, Methane, Nitrous, Oxide, Water, Vapour, and Gas.

'Means?' asked Big Wolf finally. He immediately regretted his word choice. Means made him think of mince, and now he wanted some freshly minced meat.

'That which we call a rose by any other name would smell as sweet, said William Shakespeare,' Gretel explained.

'Now who is William Shakespeare?' Big Wolf was getting tired of more and more people popping out of the blue and giving tales of woe. 'Is his rose garden affected by this climate?'

Hansel shook his head sadly and said, 'It means that names don't matter, but your deeds and characters do.'

'Of course, names matter,' Fairy Nani said. 'Names have power, silly boy.'

ENOUGH

'It's not like you have a name,' SB pointed out, reasonably. 'Everyone just calls you Fairy Nani.'

'Enough,' a voice thundered. Big Wolf turned yet again. This time, it was the Snow Queen. She seemed to have finally gathered herself. Her clothes were no longer sopping wet, but were as stiff as a school principal's spine. Her gaze was back to piercing icy blue, even though she was still a bit melty around the edges.

'Stop squabbling like children,' she said.

Hansel cleared his throat.

'Fine, fine, stop squabbling like grown-ups. It's very unbecoming. Fairyland is in the middle of its worst crisis ever, and all you can do is spread malicious rumours and quarrel among yourselves.' Her tone was getting colder and crisper by the minute, much to everyone's relief. 'Now, we have to do something. And these children here, they say that…'

'But why,' came a voice. Big Wolf sighed and craned his neck. Ouch, that whooshing in his ears had got worse. Anyway, who was that, asking such a silly question. 'But why do we have to do anything?'

The crowd gasped. It was Prince Happily Ever After. He rode on a dashing motorbike and waved at everyone, even though no one was waving back at him.

CLICK!

He slid off the bike gracefully, not a hair out of place. He smirked. 'There's nothing to be done,' he said, silkily. He walked up to the princesses and flicked his silky hair. Cinderella scowled, while No White and SB looked startled at his familiarity.

'Just look around you—what was once a silly, snow-filled land is now full of heat. We can grow things here, we can grow grapes, make wine! We can grow roses.' He winked at the princesses, who rolled their eyes in response.

'But what about the forests?' asked Big Wolf.

'We don't need them when we can get palm oil. We need land to build more factories. Factories, not forests.' He waved at No White's Seven Friends, who also look startled. 'Concrete, not jungle.'

'And what about the air? It's disgusting!' asked SB.

'I don't find it disgusting,' Prince Happily Ever After said. He took Fairy Nani's wand and stared at his reflection in it, beaming. 'It's air, it changes. Nothing is happening, fairy people. This is all nature!'

'There is definitely nothing natural about this,' the Snow Queen said, her voice reverberating like thunder. 'How dare you suggest otherwise!'

Everyone winced. Everyone except Prince Happily Ever After. 'Oh please, you all need to lighten up.'

'How can you say that!' Cinderella asked. 'We haven't been able to find Rapunzel because her tower has been enveloped by so much smog that we can't locate it.'

No White jumped. 'What, we don't know where Rapunzel is!? Oh, that's no good. She lives so high up, and the smog will affect her first.' She went up to Cinderella and held out her hands. Cinderella stared at them.

'Go on,' whispered Big Wolf.

'Go on what?' she whispered back.

'Take her hands.'

'But why? That's just bizarre. Who knows when she last washed them.'

'She's offering a hand in friendship.'

'But it's two hands she's offering.'

'OK, she's offering two hands in friendship. Just take them.'

Cinderella squirted hand sanitiser and held No White's hands.

'Forgive me, oh do, Cinderella,' No White twittered. And Cinderella immediately melted. Mind you, it may have been the heat. But still. She nodded and squeezed No White's hands back.

Meanwhile Prince Happily Ever After was still holding court. 'Rapunzel is probably fine. You all are just overreacting. Here, let me call her.' He pulled out his phone and tried calling her.

The number you are calling is enveloped in smog. Please clear the air and try again. Aap jis number ko call kar rahe hain woh filhaal smog mein gum gaya hai. Krupya hawa ki safai karein aur dobara koshish karein.

'Oh well,' he said, putting his phone back, after taking a selfie.

The next thing Prince Happily Ever After knew was that he had been hoisted onto his steed again. No White, Cinderella, Chameli, and SB had picked him up and tossed him on to his bike.

'Wha… hey stop, listen. I am right,' he yelled, as he tried to upright himself and hold the handles. Fairy Nani swished her wand, covering him with fairy dust. That was the last (paper) straw. The prince revved up his bike and sped away. He would need to make an appointment with the beauty parlour to rinse out that sticky fairy dust.

The princesses laughed and exchanged more high-fives and fist-bumps. But then they quickly sobered up. Rapunzel needed help. Fairyland was in danger. 'What can we do?' they asked in unison. They looked at each other and said 'Jinx!'

Then, as one, the Fairyland people turned to the Snow Queen and Gretel and Hansel.

7

A VERY GRIMM REALITY

'We've got to go to the end of our world,' Gretel said, looking at them as if it was really obvious. A gasp went around the crowd. No one had ever been to the End of the World; it was unheard of. No one even knew where it was.

Now, the thing is, the human world is round. Because it's housed on a planet called Earth. The fairy tale world, however, is flat—like the page of a book. Because... well... it is housed inside the pages of many books.

'But... but...' Big Wolf looked alarmed; his Big Eyes bigger than usual. 'It's the End of the World!'

Gretel rolled her eyes and said, 'Just the geographical end of our world. It's not like the end-end of our world. Our world never ends, not even after someone reads those last two words,' she dropped her voice to a whisper, 'The End'.

A collective sigh of relief echoed around the crowd.

'But how will we get there?' No White still could not understand. 'No one has ever gone there. No one really knows where it is. Do you?'

Gretel and Hansel looked at each other and shook their heads. Their shoulders sagged and they looked like they would cry. 'We thought someone here... maybe the Snow Queen.'

An angry ripple went around the crowd. 'So, you don't actually know anything,' a Twittering Troll grunted.

RRR and Big Wolf were staring at each other, not listening to the crowd. Big Wolf had finally realised something. He spoke up. 'But I do. I mean... I think I do know where it is. I think all of us animals know. We can sense it. Actually, we can hear it too. Can't you?' He looked around: all the fairy animals were nodding. There had been nothing wrong with his Big Ears. They had picked up a noise that he now realised was being emitted from the End of the World, warning him that something was wrong. 'I will lead,' he leapt up and began racing towards the End of the World.

The procession followed. It was a curious yet beautiful one.

First came Big Wolf and RRR. Followed by Gretel and Hansel, sitting in the Snow Queen's chariot, which was kept cool inside a massive tub of dry ice, and dragged by four glittery unicorns.

Then came the cabbage carriage, in which were squeezed Cinderella, No White and SB.

Fairy Nani, mask intact, hitched a ride with Chameli because she really needed some air, no matter the AQI. Cinderella had quickly shown Fairy Nani the carpet when she had declared she was so excited about another road trip that she could cry. Really, gratitude was missing in this climate-changed world, thought Fairy Nani morosely.

The motley crew of carriages, carts, steeds, dragons, and fairy folks—goblins and ogres, giants and trolls, witches and wizards, elves and fairies—followed, singing songs and breaking into dances every few minutes.

At last, they reached the End of the World, and gathered at the restaurant there. Sadly, it was shut for renovations, so Big Wolf could not get himself any food. But he cheered up when Mer popped up in the Ocean at the End of the World, right alongside a pair of unicorn whales.

'Now what?' Cinderella said, looking around her.

end, and he began growling.

'Shhh…' Hansel said, again.

Gretel turned to him and said, 'It's what we suspected.' She didn't look happy at being right. Instead, she looked downright miserable as if Nice, Nasty Witch was going to gobble her right there and then.

'What… exactly did you suspect?' the Snow Queen asked. She also looked numb. But then, given she was the Snow Queen, she always did look a bit numb.

'That the Human World is polluting ours,' Gretel said, sadly. A gasp echoed through the crowd. They looked back into the ebony nothingness and the rip reappeared, pumping grey, noxious smoke.

'As their world changes, our stories also change. Which is why where once stood mighty forests—'

'It's now full of palm oil cultivation, and so much concrete!' Big Wolf said. 'It all makes sense now. The factories, the mines and quarries that have suddenly sprung open in Fairyland. And clearly, enterprising princes and princesses have begun profiting from that. After all, stories evolve with time. Sometimes they become better, like Cinderella and SB not needing rescuing by a prince. They discovered they could rescue themselves from tricky situations.

But this time, our stories took a turn for the worse.'

'Isn't it obvious,' Hansel said, softly. 'Every time someone opens a book in the Human World, a bit of our world steps into theirs.'

Everyone nodded. That was indeed a fact. That is how stories became real. It was truly a thing of wonder.

'But now their world has changed so much. It's so full of horrible things—of smog and dread, of fear and noxious gases, of pandemics and trash—that it has nowhere else left to go.'

'Except, into our world,' Cinderella said. 'It became so big there, that it tore into our land.'

Everyone was so shocked that they forgot to gasp.

The two teenagers nodded unhappily. They looked at the crowd. Everyone looked as grim as the Brothers Grimm were rumoured to be.

And then Gretel and Hansel held hands, and it spread across the fairy people. Everyone held each other's hands, even Cinderella (after a squirt of sanitiser). They all closed their eyes.

8

THE OTHER WORLD

Once upon a time, a child opened a book.
Another child read a book.
Yet another heard as her grandmother
told her a story from a book.
One child heard it in school.
Many heard it in a library.
As did another.
And another.

All of them stepped into another world, where they met carriages, carts, steeds, dragons and fairy folks, goblins and dwarves, giants and trolls, witches and wizards, elves and fairies. A world where Snow White sang in the forest sprinkled with her namesake, and Cinderella danced without her name being blackened

by sooty accusations, where Big Bad Wolf lived in the Enchanted Forest, alongside the Nice, Nasty Witch, Gretel and Hansel, and Rapunzel's tower was visible under a clear blue sky, where Little Mermaid's biggest problem was a silly prince. Stories so familiar. Retold, recast, made anew with every retelling.

But then, most of these children grew up and forgot what they had read in those books. That magic is real.

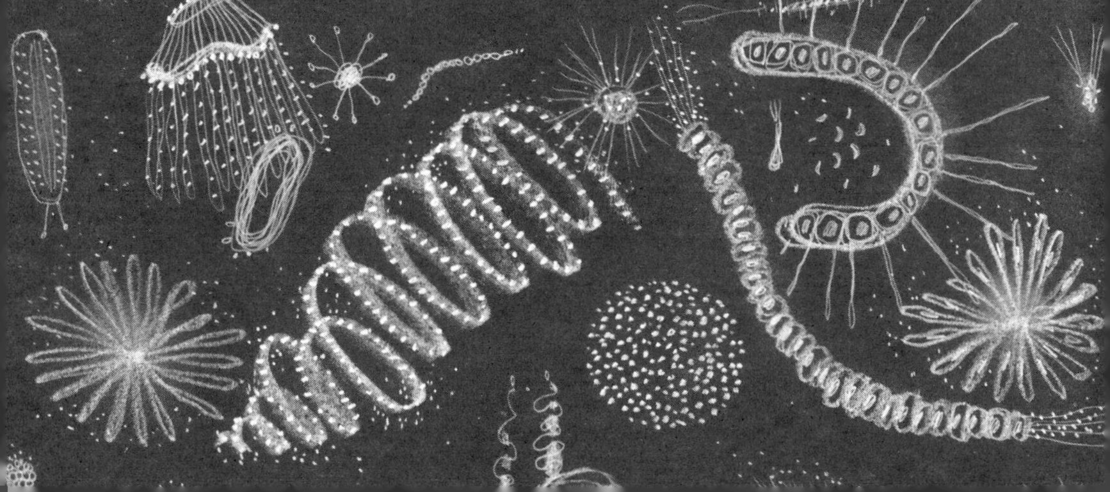

That new worlds can be found on top of trees
and under toadstools.
The shimmering fireflies dance like fairies lighting
our lives like moonlight.
That all frogs are charming and all lions and lionesses, kings.
That a butterfly is a fairyland creature, magical in its
transformation and powers of pollination.
That the turning of seasons is an enchantment in itself.
That phytoplankton are magicians, casting a blue
bioluminescence on waves.
That the song of a red-vented bulbul gladdens the weary heart.
That the pied cuckoo holds a secret in her heart, which tells
her when the monsoon will visit.
That, by the turn of a head or the closing of a wing, bats
and owls step into another realm.
That spiders are sorceresses who spin stories.
And that the Earth is the witch and the wizard, in whom reside
many powers—of little seeds that magic themselves
into the tallest of trees and conjure up clean air, of snow
and ice who guard water, of dung beetles
and bacteria that clean up after everyone.

As the children grew up, they forgot all of this.

As grown-ups, they forgot about magic, and stopped believing in the magic of nature. They forgot that magic needs to be nurtured and celebrated, not conquered or controlled. They dug through mountains and cut up the earth's heart for coal and metal; they ploughed down forests to make way for smoky factories; they silenced the rivers with dams; they built upon grasslands and punctured the earth's lungs to punctuate their greed; they consumed and consumed and chucked and chucked, until lands became landfills.

9

HAPPILY EVER AFTER

Gretel opened her eyes. As did Hansel. As did all the other fairy people. It was like stepping out of a dream.

'What do we do now?'

'What is the point of seeing this?'

'Arre yaar!'

'Need payasam!'

The last one was from Little Bear, whose mother quickly gave him a slice of beehive to suck on. She'd had to raid her emergency frozen stack ever since the bees disappeared.

'Now,' Gretel said. 'We do what Fairyland is known for doing.' The sister and brother turned to the Snow Queen, who looked back at them.

Big Wolf really wished he could read thoughts because he couldn't figure out what they were thinking.

The Snow Queen smiled, and it was like the first snowfall of the season. Everyone's hearts felt a little lighter, gave a little skip, as if ready to go out and dance in the flurries of fresh snow. 'We bestow gifts,' she said.

'Gifts?' SB said. 'What gifts? I'd like a gift.'

The Snow Queen pretended not to hear SB and continued. 'Isn't that what we do? We gift—the gift of songs, of beauty, of wit, of goodness…'

'Of pumpkins, I mean, cabbages to carriages!' added Fairy Nani.

'Of friendship,' No White held Cinderella's hand again. Cinderella looked like she was going to snatch it away, but she didn't.

'Also curses,' Nice, Nasty Witch pointed out.

The Snow Queen frowned and then nodded, 'Yes, what are gifts without curses? We wouldn't appreciate a rainbow without a storm, a silver lining without a black cloud. You speak wisely.'

'Excuse me,' a voice piped up. This time it was Big Wolf. Everyone turned to look at him. 'I'm sorry, but why are we giving them gifts again? They're wrecking our world, and we're gifting them things.

What is this? A "burn 3 fossil fuels, get 20 gifts" sale?' He was honestly fed up. Fed up with the hunger in his Big Belly; tired of living in a Disenchanted Not-a-Forest; sick of all the horrible things that were happening to all of them.

'We must develop and maintain the capacity to forgive. He who is devoid of the power to forgive is devoid of the power to love,' Gretel said.

'Now who said that?' Big Wolf asked. 'That Will fellow?'

Gretel rolled her eyes. 'It was Martin Luther King Jr.'

'Dunno what they teach you in those schools,' Big Wolf growled, while reluctantly realising the girl was right. He pulled a face at her and shrugged. 'Fine, let's have it your way. Anything to make this go away.'

The Snow Queen nodded and in front of her stepped—

Nice, Nasty Witch. 'I bestow upon the Human World the gift of renewed sight, smell, taste, touch, sound, and feeling so they can explore and respect the natural world,' she said. Something silvery flew out of her cauldron into the rip.

'I bestow upon the Human World the gift of colour. May the corals return to their rainbows, the gurgles to the rivers, and may the oceans be filled with colourful life and not plastic,' Mer said, her tail creating a whirlpool, which sent splashes into the rip.

'We bestow upon the Human World a sense of wonder, like every child is born with, to sustain the joy that their world is,' Gretel and Hansel said. They high-fived, its echo travelling into the ebony nothingness.

'We bestow upon the Human World a friendship that brings together the collective wisdom of the people who live with the land,' the princesses said in unison. A trail of flowers and moths flowed into the rip.

'We bestow upon the Human World a regard for the gloriousness of seasons, of harvests and blossoms,' growled Bear family. Mama Bear removed a dandelion from her pocket and blew on it, sending the florets gently flying into the rip, which was getting smaller.

'And lots of payasam,' Goldilocks quickly added. A glob of payasam followed suit.

RRR stepped out in front.

'I bestow upon the Human World
(Who has behaved like a turd)
An understanding of forest and tree
A respect of flora, fauna; a mind that's free.'

Their poem travelled into the hole.

RRR turned to Big Wolf and shook their head at him. Big Wolf sighed loudly and said, 'FFIIIINE!' He stepped in

front and rumbled, 'I bestow upon you an appetite for knowledge, instead of greed.' Golden light danced into the rip. 'Happy?' he said to the crowd. But when everyone broke out into applause, he couldn't help but smile. Okay, maybe being nice wasn't that bad.

Finally, it was the Snow Queen's turn. 'I bestow upon the Human World the gift of snow and wisdom, may their polar caps not melt further, and may the wisdom of generations of ice stories spread to them,' she said. Snow bees burst from her palm, and squeezed into the rip, which was now the size of a keyhole.

And before anyone could stop her, Fairy Nani waved her wand and trilled, 'I bestow upon the Human World fairy dust.'

Everyone groaned.

'Nooooo,' Page 89 wailed, who doubled as carriage keeper.

The Snow Queen held out her hand to try and stop the dust. But it was too late, and the fairy dust had travelled into the Human World.

'Oh well, no one really lives happily ever after,' the Nice, Nasty Witch shrugged.

The End

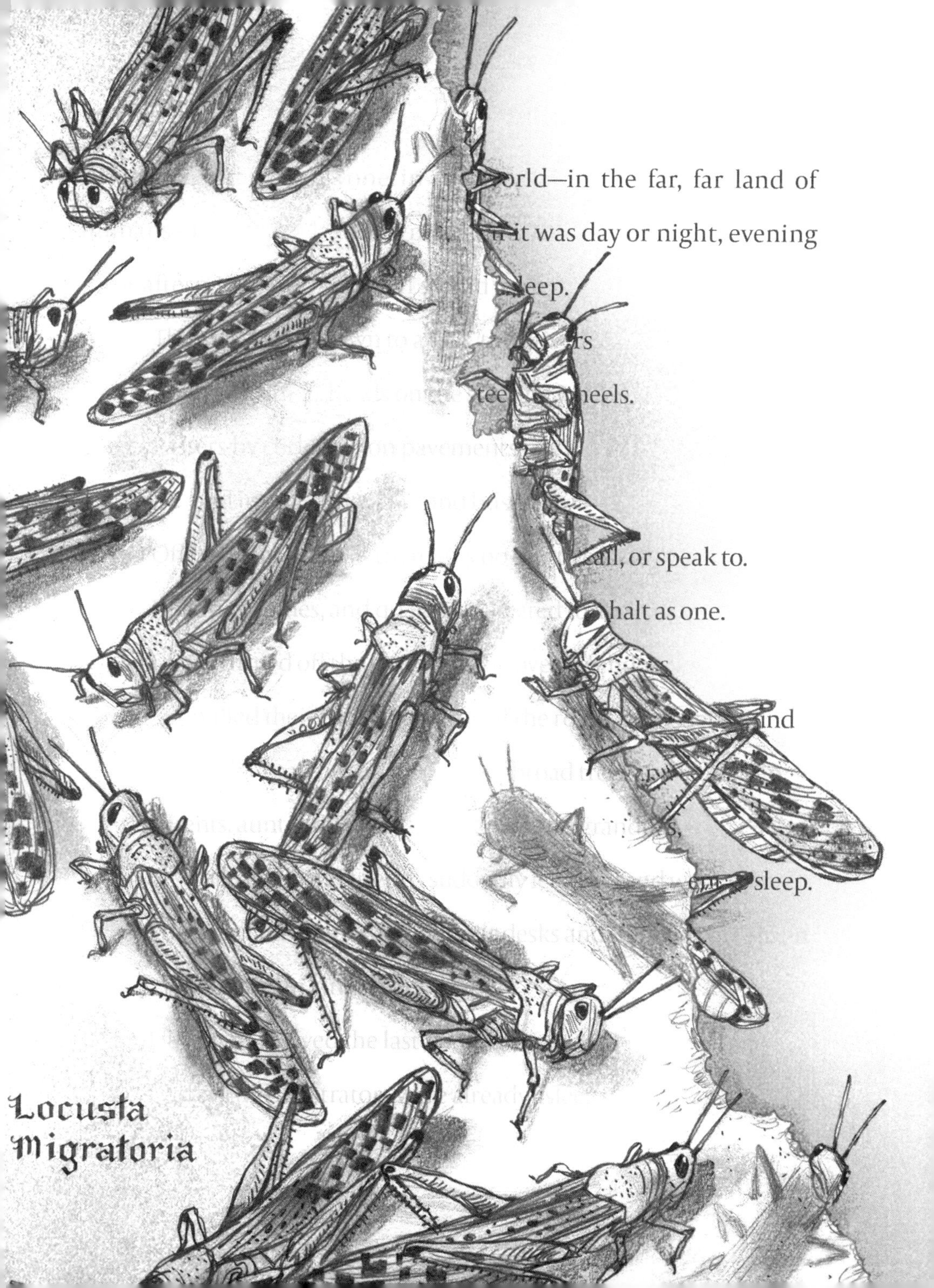
orld—in the far, far land of
it was day or night, evening
leep.
rs
heels.
all, or speak to.
halt as one.
ind
sleep.
Locusta
Migratoria

IN A FAR, FAR LAND

In a land far, far away, sat a little girl. She was wide awake and trying hard to remember her favourite story, to try and go back to sleep. She knew the story so well, but it was just out of reach. Like chasing a butterfly on a dazzling sun-drenched day.

The little girl opened her favourite story and gasped as words started to appear. They weren't from the story she knew so well. They told a different story.

She read it, wide-eyed. Read it until she knew it by heart. She read it until she could no longer smell its pages. She read it until its touch felt like her parents' touch on her forehead—familiar and comforting.

She read until the words disappeared from the pages.

But this time, she didn't worry. Instead, the girl finally drifted off to sleep.

Just like that, everyone in the world—in the far, far land of humans—fell asleep. It didn't matter if it was day or night, evening or afternoon—everyone fell sound asleep.

The cars slowed down to a halt, the drivers
resting their heads on the steering wheels.
Passers-by curled up on pavements,
and in auto rickshaws and buses.
Offices went silent as there was no one to call, or speak to.
Factories, mines, and quarries clattered to a halt as one.
Cooks turned off the gases and the ovens, vendors
pulled their carts to the side of the roads, gardeners found
patches of grass under broad trees and fell asleep.
Parents, aunts and uncles, grandmas and grandpas,
colleagues and friends suddenly felt tired and went to sleep.
Students laid their heads on their desks and stepped
into dream land.
Librarians shelved the last book.
Writers and illustrators were already asleep.
And they dreamed.
Of a better world.
Where the magic that is nature was threaded
into a happily ever after.

Not the End

We thank thee, Book Makers

Robert Macfarlane—Tortoiseshell Hare—dropped a bowl of alphabet soup at the Wordsmith Café. He quickly invoked a lost spell. Wherever the alphabets fell, grew a forest anew. Now wherever he goes, much to his horror, people hand him bowls of too-hot soup.

Zai Whitaker—Prickly Pinocchio—went splash into the moat. An Alley-Gator swam up to her and swallowed her whole. In that tummy, she lived happily ever after, with a king cobra and a termite hill. Except once a year, when she would go to see her dentist.

Rohan Chakravarty—the Green Humerus Hag—tried out a new spell. 'A pangolin I'll be', he said. 'Pholidota!' The spell went just a bit wrong. He was doomed to belt out Bollywood songs to anyone who asked him the time of day.

Aparna Kapur—the Small Unfriendly Giant (SMUG)—sculpted the most fabled stories with a hammer stolen from Thor. She courageously fought the Kadoo Beast in the Gully of the Storye Swamp. And reached home, just in time to save the tub of mulberry ice cream from melting.

Deepanjana Pal—Djinnius—was bestowed with the gift of turning all tall people invisible. She went to the Towering Library to spar with the Very Tall Librarian. She sat down to read a book, then read another, and another, and hasn't been seen since.

Maithili Doshi—the Never Hobnobbing Goblin—sent out an incantation; the sentences arranged themselves, the paragraphs ran themselves right, and the pages paginated themselves. She then set out to duel a deadline.

Manojna Yeluri—the Banishing Banshee—wailed until the Words put on fancy head phones. When noise cancellation didn't work, they had no choice but to behave themselves.

Kanishka Gupta—the Where Wolf—went from town to town talking about a mythical codex. 'Hear, hear, who wants this codex?' he asked with a sigh. 'I do' said Riding Hood, and swallowed the Wolf, codex and all.

Smit Zaveri—the Towering Teacup—baked S'more Sentences, Conjunction Cakes, and Present Tense Puddings. She baked up a storm in a teacup, until it became a tornado. Now she's best friends with Dorothy and Toto.

Tina Narang—the Publishing Fairy—got lost in a labyrinth. With a map and a ball of yarn, she found her way out to the literary trenches of HarperCollins. Oh no, Anita Sharma, Garima Syal, Nisha Singh, Radhika Shenoy and Nimmy Chacko were still in the labyrinth.

BIJAL VACHHARAJANI—Gruffin' Griffin—found herself sneezing her head off at fairy dust and the rising AQI (Air Quality Index). She needed her head to write more planet-friendly books and worry about the climate crisis. But now her head is rolling down a hill, and she needs to chase it before it becomes part of a nursery rhyme, or worse, a reel on social media.

RAJIV EIPE—Creature of Hobbiton—was aghast at the rising temperature of planet Earth. It was so hot that his sketchbook transformed into a squelchbook. His sketches became messes, his doodles turned into puddles of poodles. Now, he can't even drown his sorrow in a ginormous cup of coffee because it's too hot to drink it.

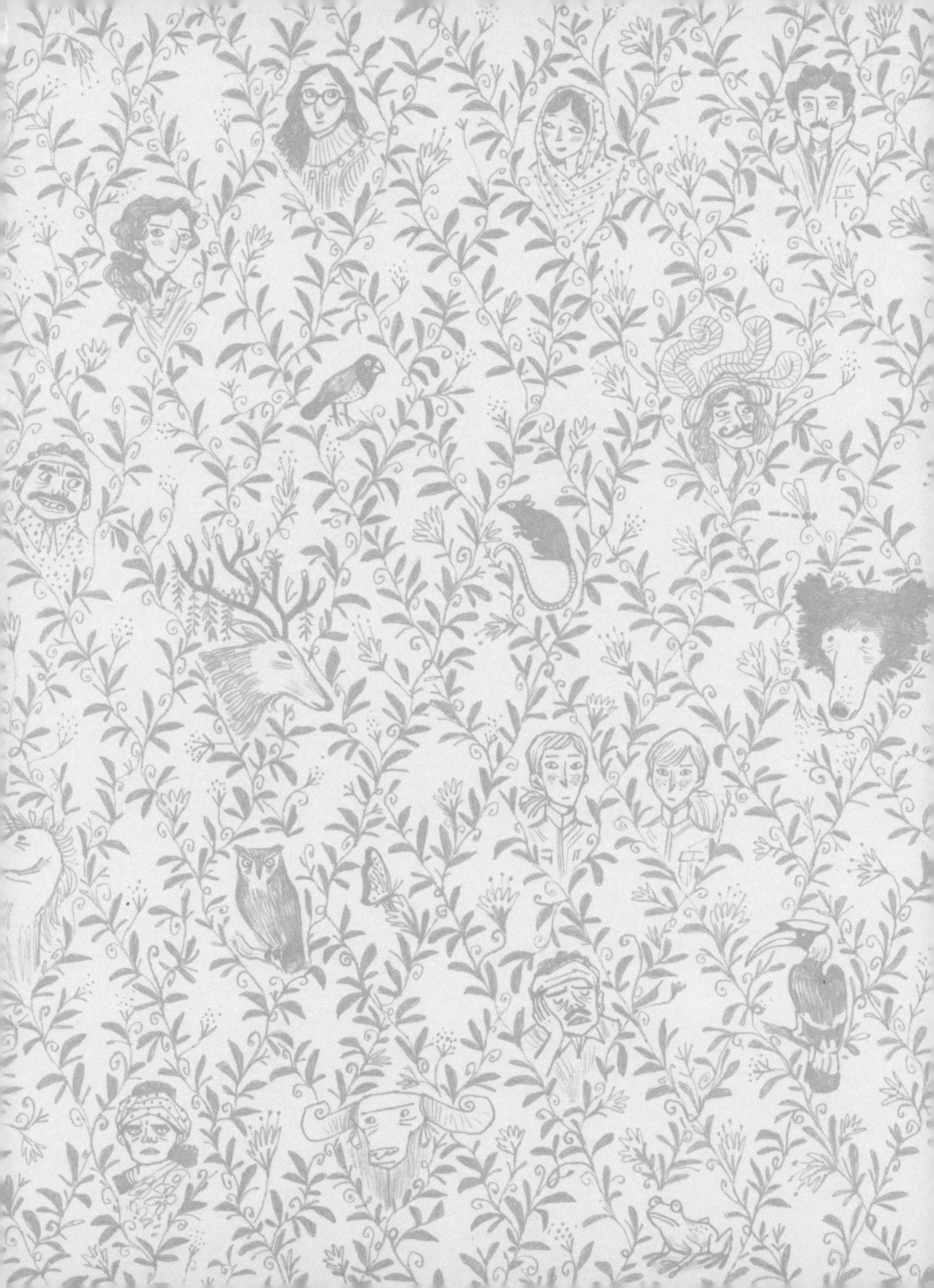